Math Games LEVEL D
Centers for Up to 6 Players

How to Use Math Games 2
How to Make a Math Game 3
Math Games Checklist 4

Game 1 **Concentration** *Addition and Subtraction Review*

Game 2 **Multiply It** *Multiplication Facts to 12*

Game 3 **Divide It** *Division Facts to 12*

Game 4 **Which Is It?** *Less Than, Greater Than, Equal To*

Game 5 **Ping Pang Pow!** *Money: Equal Amounts*

Game 6 **What Time Is It?** *Time to the Minute*

Game 7 **Concentration** *Equivalent Fractions*

Editorial Development: Joy Evans
Jo Ellen Moore
Lisa Vitarisi Mathews
Copy Editing: Carrie Gwynne
Art Direction: Cheryl Puckett
Cover Design: Liliana Potigian
Design/Production: Marcia Smith
Olivia C. Trinidad

EMC 3032

Congratulations on your purchase of some of the finest teaching materials in the world.

Photocopying the pages in this book is permitted for single-classroom use only. Making photocopies for additional classes or schools is prohibited.

For information about other Evan-Moor products, call 1-800-777-4362, fax 1-800-777-4332, or visit our Web site, www.evan-moor.com. Entire contents © 2008 EVAN-MOOR CORP. 18 Lower Ragsdale Drive, Monterey, CA 93940-5746. Printed in USA.

Correlated to State Standards

Visit *teaching-standards.com* to view a correlation of this book's activities to your state's standards. This is a free service.

How to Use Math Games

Play the games as a follow-up to a math lesson, or use a game to target a skill that several students need to practice. The games are also fun "extra-time" or rainy-day recess activities.

Model how to play each type of game, and place the games in an area of your classroom that is easily accessible to students.

Games Include:

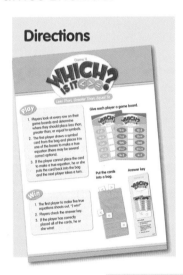

Directions

Game boards

Game cards

Answer key

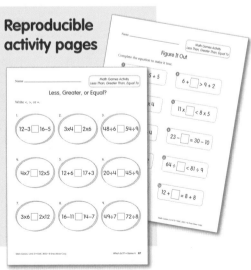

Reproducible activity pages

How to Make a Math Game

Steps to Follow:

1. Laminate the directions page, the game boards, the cards, and the answer key(s).

2. Reproduce the activity pages.

3. Place the laminated game supplies and any additional items, such as bean markers or brown paper bags, in a folder that has a closure.

Materials

- laminator
- scissors
- brown paper bags
- game board markers such as beans
- folder that has a closure

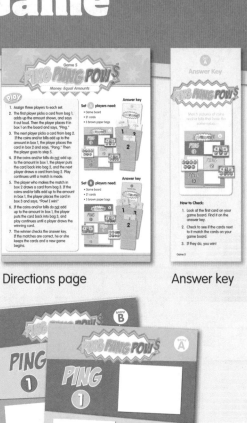
Directions page · Answer key

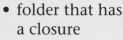

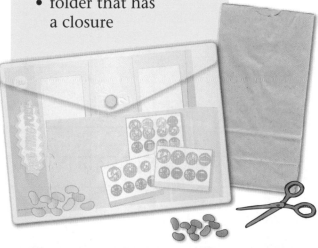

Game boards · Game cards · Reproducible activity pages

EMC 3032 • © Evan-Moor Corp.   Math Games • Level D   3

Math Games Checklist

LEVEL D

Student	Games Played	Concentration Addition and Subtraction Review	Multiply It Multiplication Facts to 12	Divide It Division Facts to 12	Which Is It? Less Than, Greater Than, Equal To	Ping Pang Pow! Money: Equal Amounts	What Time Is It? Time to the Minute	Concentration Equivalent Fractions

Game 1

Concentration

Addition and Subtraction Review

1. Place each set of cards facedown in six rows of four. Assign three players to each set.
2. The first player in each group turns over an addition or subtraction card and an answer card. If the answer card solves the addition or subtraction problem, the player keeps the pair of cards and plays again.
3. If the answer card does <u>not</u> solve the addition or subtraction problem, the player turns the cards over and the next player takes a turn.

Set A players need:
- 24 purple cards

Answer key

Set B players need:
- 24 green cards

Answer key

1. Play continues until all of the cards are matched.
2. Then each player looks at the answer key to make sure his or her cards are correctly matched.
3. The player with the most pairs wins!

9+6	11+3	7+5
18+2	13+7	4+8
10+5	14+3	8+9

Game 1 • Set A

Math Games, Level D
© Evan-Moor Corp.

Game 1 • Set A

Math Games, Level D
© Evan-Moor Corp.

Game 1 • Set A

Math Games, Level D
© Evan-Moor Corp.

Game 1 • Set A

Math Games, Level D
© Evan-Moor Corp.

Game 1 • Set A

Math Games, Level D
© Evan-Moor Corp.

Game 1 • Set A

Math Games, Level D
© Evan-Moor Corp.

Game 1 • Set A

Math Games, Level D
© Evan-Moor Corp.

Game 1 • Set A

Math Games, Level D
© Evan-Moor Corp.

Game 1 • Set A

Math Games, Level D
© Evan-Moor Corp.

12+7	11+8	8+6
15	14	12
20	20	12

Game 1 • Set A

Math Games, Level D
© Evan-Moor Corp.

Game 1 • Set A

Math Games, Level D
© Evan-Moor Corp.

Game 1 • Set A

Math Games, Level D
© Evan-Moor Corp.

Game 1 • Set A

Math Games, Level D
© Evan-Moor Corp.

Game 1 • Set A

Math Games, Level D
© Evan-Moor Corp.

Game 1 • Set A

Math Games, Level D
© Evan-Moor Corp.

Game 1 • Set A

Math Games, Level D
© Evan-Moor Corp.

Game 1 • Set A

Math Games, Level D
© Evan-Moor Corp.

Game 1 • Set A

Math Games, Level D
© Evan-Moor Corp.

15	17	17
19	19	14
20-7	14-11	17-4

Game 1 • Set A

Math Games, Level D
© Evan-Moor Corp.

Game 1 • Set A

Math Games, Level D
© Evan-Moor Corp.

Game 1 • Set A

Math Games, Level D
© Evan-Moor Corp.

Game 1 • Set A

Math Games, Level D
© Evan-Moor Corp.

Game 1 • Set A

Math Games, Level D
© Evan-Moor Corp.

Game 1 • Set A

Math Games, Level D
© Evan-Moor Corp.

Game 1 • Set B

Math Games, Level D
© Evan-Moor Corp.

Game 1 • Set B

Math Games, Level D
© Evan-Moor Corp.

Game 1 • Set B

Math Games, Level D
© Evan-Moor Corp.

16-5	19-8	20-17
12-8	16-12	15-3
20-8	12-7	18-13

Game 1 • Set B

Math Games, Level D
© Evan-Moor Corp.

Game 1 • Set B

Math Games, Level D
© Evan-Moor Corp.

Game 1 • Set B

Math Games, Level D
© Evan-Moor Corp.

Game 1 • Set B

Math Games, Level D
© Evan-Moor Corp.

Game 1 • Set B

Math Games, Level D
© Evan-Moor Corp.

Game 1 • Set B

Math Games, Level D
© Evan-Moor Corp.

Game 1 • Set B

Math Games, Level D
© Evan-Moor Corp.

Game 1 • Set B

Math Games, Level D
© Evan-Moor Corp.

Game 1 • Set B

Math Games, Level D
© Evan-Moor Corp.

13	3	13
11	11	3
4	4	12

Game 1 • Set B

Math Games, Level D
© Evan-Moor Corp.

Game 1 • Set B

Math Games, Level D
© Evan-Moor Corp.

Game 1 • Set B

Math Games, Level D
© Evan-Moor Corp.

Game 1 • Set B

Math Games, Level D
© Evan-Moor Corp.

Game 1 • Set B

Math Games, Level D
© Evan-Moor Corp.

Game 1 • Set B

Math Games, Level D
© Evan-Moor Corp.

Game 1 • Set B

Math Games, Level D
© Evan-Moor Corp.

Game 1 • Set B

Math Games, Level D
© Evan-Moor Corp.

Game 1 • Set B

Math Games, Level D
© Evan-Moor Corp.

| 12 | 5 | 5 |

Game 1 • Set B

Math Games, Level D
© Evan-Moor Corp.

Game 1 • Set B

Math Games, Level D
© Evan-Moor Corp.

Game 1 • Set B

Math Games, Level D
© Evan-Moor Corp.

Answer Key

Match each addition problem to its answer card.

How to Check:

1. Look at one of your pairs. Look at the addition problem. Find the same addition problem on the answer key.

2. See if the card next to it matches your answer card. If it does, you made a pair.

3. Check your other pairs.

Game 1

Answer Key

Match each subtraction problem to its answer card.

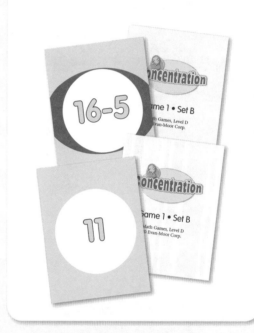

How to Check:

1. Look at one of your pairs. Look at the subtraction problem. Find the same subtraction problem on the answer key.

2. See if the card next to it matches your answer card. If it does, you made a pair.

3. Check your other pairs.

Game 1

Set B

20-17	=	3
14-11	=	3
12-8	=	4
16-12	=	4
12-7	=	5
18-13	=	5
19-8	=	11
16-5	=	11
20-8	=	12
15-3	=	12
20-7	=	13
17-4	=	13

Set A

4+8	=	12
7+5	=	12
11+3	=	14
8+6	=	14
10+5	=	15
9+6	=	15
8+9	=	17
14+3	=	17
11+8	=	19
12+7	=	19
13+7	=	20
18+2	=	20

Math Games, Level D
EMC 3032 • © Evan-Moor Corp.

Name _____

Math Games Activity
Addition Review

Fun Facts Addition

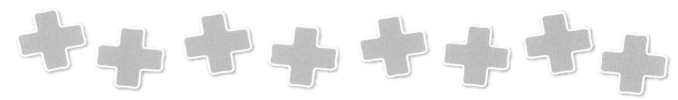

Add. Write the sum.

```
  16      11       9       9       8
+  2     + 4     + 2     + 9     + 3


  12       4       5       8      12
+  7     +13     + 5     + 7     + 4


   5       6       9      13      16
+  3     + 5     + 4     + 6     + 2


  17       9       8      10       7
+  3     + 6     + 5     + 3     + 4
```

Name _____

Math Games Activity
Subtraction Review

Fun Facts Subtraction

Subtract. Write the difference.

```
  13        9       12       17       20
-  4     -  3     -  6     -  5     -  7
____     ____     ____     ____     ____


  11        8       14        6        9
-  3     -  4     -  9     -  4     -  2
____     ____     ____     ____     ____


  18       14        7       12       11
-  9     -  6     -  5     - 10     -  6
____     ____     ____     ____     ____


  18       20       19       13       17
- 11     - 14     - 10     -  8     - 13
____     ____     ____     ____     ____
```

Game 2
Multiply x It

Multiplication Facts to 12

1. The first player picks a card from the bag and reads the number out loud.
2. If the number solves a multiplication problem on the player's board, the player places the number card beside it. The player draws again.
3. If there is already a card beside the problem, or the number does <u>not</u> solve a problem, the player puts the card back into the bag.
4. The next player takes a turn.

Give each player a game board.

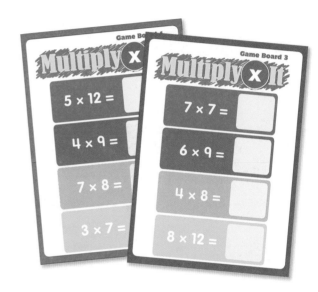

1. The first player to place cards beside all four multiplication problems shouts out, "I win!"
2. Players check the answer key.
3. If the player has correctly solved all four problems, he or she wins!

Put the cards into a bag.

Answer key

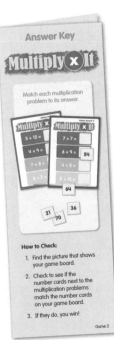

Game Board 1

Multiply x It

5 × 12 =

4 × 9 =

7 × 8 =

3 × 7 =

Math Games, Level D
© Evan-Moor Corp.

Game Board 2

8 × 8 =

4 × 12 =

7 × 10 =

2 × 12 =

Math Games, Level D
© Evan-Moor Corp.

Game Board 3

7 × 7 =

6 × 9 =

4 × 8 =

8 × 12 =

Math Games, Level D
© Evan-Moor Corp.

Game Board 4

9 × 11 =

4 × 12 =

8 × 8 =

7 × 9 =

Math Games, Level D
© Evan-Moor Corp.

Game Board 5

5 × 5 =

3 × 8 =

9 × 10 =

5 × 11 =

Math Games, Level D
© Evan-Moor Corp.

Game Board 6

4 × 11 =

5 × 7 =

3 × 9 =

8 × 12 =

Math Games, Level D
© Evan-Moor Corp.

60	36	56
21	64	48
70	24	49
54	32	96
99	48	64

Math Games, Level D
© Evan-Moor Corp.

Math Games, Level D
© Evan-Moor Corp.

Math Games, Level D
© Evan-Moor Corp.

Math Games, Level D
© Evan-Moor Corp.

Math Games, Level D
© Evan-Moor Corp.

Math Games, Level D
© Evan-Moor Corp.

Math Games, Level D
© Evan-Moor Corp.

Math Games, Level D
© Evan-Moor Corp.

Math Games, Level D
© Evan-Moor Corp.

Math Games, Level D
© Evan-Moor Corp.

Math Games, Level D
© Evan-Moor Corp.

Math Games, Level D
© Evan-Moor Corp.

Math Games, Level D
© Evan-Moor Corp.

Math Games, Level D
© Evan-Moor Corp.

Math Games, Level D
© Evan-Moor Corp.

63	25	24
90	55	44
35	27	96
72	60	84
55	67	43

Math Games, Level D
© Evan-Moor Corp.

Math Games, Level D
© Evan-Moor Corp.

Math Games, Level D
© Evan-Moor Corp.

Math Games, Level D
© Evan-Moor Corp.

Math Games, Level D
© Evan-Moor Corp.

Math Games, Level D
© Evan-Moor Corp.

Math Games, Level D
© Evan-Moor Corp.

Math Games, Level D
© Evan-Moor Corp.

Math Games, Level D
© Evan-Moor Corp.

Math Games, Level D
© Evan-Moor Corp.

Math Games, Level D
© Evan-Moor Corp.

Math Games, Level D
© Evan-Moor Corp.

Math Games, Level D
© Evan-Moor Corp.

Math Games, Level D
© Evan-Moor Corp.

Math Games, Level D
© Evan-Moor Corp.

Answer Key

Multiply x It

Match each multiplication problem to its answer.

How to Check:

1. Find the picture that shows your game board.

2. Check to see if the number cards next to the multiplication problems match the number cards on your game board.

3. If they do, you win!

Multiply It

❶ Game Board 1
- $5 \times 12 =$ **60**
- $4 \times 9 =$ **36**
- $7 \times 8 =$ **56**
- $3 \times 7 =$ **21**

❷ Game Board 2
- $8 \times 8 =$ **64**
- $4 \times 12 =$ **48**
- $7 \times 10 =$ **70**
- $2 \times 12 =$ **24**

❸ Game Board 3
- $7 \times 7 =$ **49**
- $6 \times 9 =$ **54**
- $4 \times 8 =$ **32**
- $8 \times 12 =$ **96**

❹ Game Board 4
- $9 \times 11 =$ **99**
- $4 \times 12 =$ **48**
- $8 \times 8 =$ **64**
- $7 \times 9 =$ **63**

❺ Game Board 5
- $5 \times 5 =$ **25**
- $3 \times 8 =$ **24**
- $9 \times 10 =$ **90**
- $5 \times 11 =$ **55**

❻ Game Board 6
- $4 \times 11 =$ **44**
- $5 \times 7 =$ **35**
- $3 \times 9 =$ **27**
- $8 \times 12 =$ **96**

Name _____

Math Games Activity
Multiplication Facts to 12

Fun Facts Multiplication

Multiply. Write the product.

```
   4       10        8       12        9
 x 6      x 9      x 7      x 3      x 5
 ___      ___      ___      ___      ___


  11        5        6       10        8
 x 9      x 3      x 6      x 7      x 9
 ___      ___      ___      ___      ___


   7        3        9       12        6
 x 7      x 4      x 2      x 8      x 8
 ___      ___      ___      ___      ___


  11       12        7        6        8
 x 4      x 7      x 5      x 6      x 8
 ___      ___      ___      ___      ___
```

Name _____

Math Games Activity
Multiplication Facts to 12

Why didn't the hot dog star in the movies?

Key

6–d	16–r	21–s	30–n	56–w
10–'	18–t	22–u	36–e	
12–l	20–h	24–o	40–g	

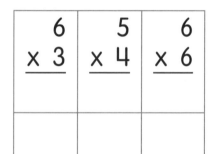

Multiply. Write the product.
Use the key to solve the riddle.

6 × 3	5 × 4	6 × 6		4 × 4	12 × 2	4 × 3	6 × 2	7 × 3

8 × 7	9 × 4	8 × 2	12 × 3	6 × 5	2 × 5	2 × 9

8 × 5	6 × 4	24 × 1	3 × 2		36 × 1	10 × 3	24 × 1	11 × 2	10 × 4	10 × 2

44 Multiply It • Game 2

Game 3

Divide ÷ It

Division Facts to 12

1. The first player picks a card from the bag and reads the number out loud.
2. If the number solves one of the division problems on the player's board, the player places the card beside the problem and draws again.
3. If there is already a card beside the problem, or the number does <u>not</u> solve a problem, the player puts the card back into the bag.
4. The next player takes a turn.

Give each player a game board.

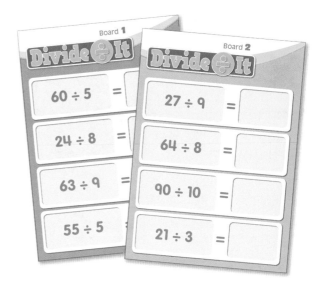

Put the cards into a bag.

Answer key

1. The first player to place cards beside all four division problems shouts out, "I win!"
2. Players check the answer key.
3. If the player has correctly solved all four problems, he or she wins!

Board **1**

Divide ÷ It

60 ÷ 5	=
24 ÷ 8	=
63 ÷ 9	=
55 ÷ 5	=

Math Games, Level D
© Evan-Moor Corp.

Board **2**

Divide ÷ It

27 ÷ 9 =	
64 ÷ 8 =	
90 ÷ 10 =	
21 ÷ 3 =	

Math Games, Level D
© Evan-Moor Corp.

Board 3

Divide ÷ It

36 ÷ 3 =

45 ÷ 9 =

56 ÷ 8 =

90 ÷ 10 =

Math Games, Level D
© Evan-Moor Corp.

Board **4**

Divide ÷ It

24 ÷ 3 =

48 ÷ 8 =

55 ÷ 5 =

96 ÷ 8 =

Math Games, Level D
© Evan-Moor Corp.

Board 5

Divide ÷ It

54 ÷ 9 =

45 ÷ 5 =

36 ÷ 3 =

70 ÷ 7 =

Math Games, Level D
© Evan-Moor Corp.

Board **6**

Divide ÷ It

72 ÷ 9	=
22 ÷ 2	=
49 ÷ 7	=
32 ÷ 8	=

Math Games, Level D
© Evan-Moor Corp.

12	3	7
11	3	8
9	7	12
5	7	9
8	6	11

Math Games • Level D
© Evan-Moor Corp.

Math Games • Level D
© Evan-Moor Corp.

Math Games • Level D
© Evan-Moor Corp.

Math Games • Level D
© Evan-Moor Corp.

Math Games • Level D
© Evan-Moor Corp.

Math Games • Level D
© Evan-Moor Corp.

Math Games • Level D
© Evan-Moor Corp.

Math Games • Level D
© Evan-Moor Corp.

Math Games • Level D
© Evan-Moor Corp.

Math Games • Level D
© Evan-Moor Corp.

Math Games • Level D
© Evan-Moor Corp.

Math Games • Level D
© Evan-Moor Corp.

Math Games • Level D
© Evan-Moor Corp.

Math Games • Level D
© Evan-Moor Corp.

Math Games • Level D
© Evan-Moor Corp.

12	6	9
12	10	8
11	7	4
12	9	7
8	8	3

Math Games • Level D
© Evan-Moor Corp.

Math Games • Level D
© Evan-Moor Corp.

Math Games • Level D
© Evan-Moor Corp.

Math Games • Level D
© Evan-Moor Corp.

Math Games • Level D
© Evan-Moor Corp.

Math Games • Level D
© Evan-Moor Corp.

Math Games • Level D
© Evan-Moor Corp.

Math Games • Level D
© Evan-Moor Corp.

Math Games • Level D
© Evan-Moor Corp.

Math Games • Level D
© Evan-Moor Corp.

Math Games • Level D
© Evan-Moor Corp.

Math Games • Level D
© Evan-Moor Corp.

Math Games • Level D
© Evan-Moor Corp.

Math Games • Level D
© Evan-Moor Corp.

Math Games • Level D
© Evan-Moor Corp.

Answer Key

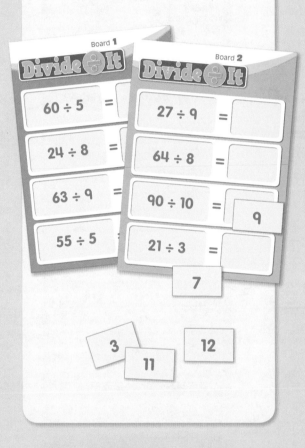

Match each division problem to its quotient.

How to Check:

1. Find the picture of your game board.

2. Check to see if the number cards next to the division problems match the number cards on your game board.

3. If they do, you win!

Divide It

❶ Board 1
- 60 ÷ 5 = 12
- 24 ÷ 8 = 3
- 63 ÷ 9 = 7
- 55 ÷ 5 = 11

❷ Board 2
- 27 ÷ 9 = 3
- 64 ÷ 8 = 8
- 90 ÷ 10 = 9
- 21 ÷ 3 = 7

❸ Board 3
- 36 ÷ 3 = 12
- 45 ÷ 9 = 5
- 56 ÷ 8 = 7
- 90 ÷ 10 = 9

❹ Board 4
- 24 ÷ 3 = 8
- 48 ÷ 8 = 6
- 55 ÷ 5 = 11
- 96 ÷ 8 = 12

❺ Board 5
- 54 ÷ 9 = 6
- 45 ÷ 5 = 9
- 36 ÷ 3 = 12
- 70 ÷ 7 = 10

❻ Board 6
- 72 ÷ 9 = 8
- 22 ÷ 2 = 11
- 49 ÷ 7 = 7
- 32 ÷ 8 = 4

Name _____

Math Games Activity
Division Facts to 12

Fun Facts Division

Divide. Write the quotient.

$$21 \div 3 \quad\quad 56 \div 8 \quad\quad 22 \div 2 \quad\quad 49 \div 7 \quad\quad 72 \div 9$$

$$16 \div 4 \quad\quad 55 \div 5 \quad\quad 36 \div 6 \quad\quad 18 \div 2 \quad\quad 54 \div 6$$

$$32 \div 8 \quad\quad 27 \div 9 \quad\quad 60 \div 6 \quad\quad 70 \div 7 \quad\quad 42 \div 7$$

$$60 \div 5 \quad\quad 54 \div 9 \quad\quad 48 \div 6 \quad\quad 45 \div 5 \quad\quad 33 \div 3$$

Name _____

Math Games Activity
Division Facts to 12

What Do You Call a Scared Dinosaur?

Solve each division problem.

A 21 ÷ 3 = _____   S 24 ÷ 2 = _____

E 24 ÷ 3 = _____   U 18 ÷ 6 = _____

N 36 ÷ 6 = _____   V 16 ÷ 4 = _____

O 54 ÷ 6 = _____   X 30 ÷ 3 = _____

R 40 ÷ 8 = _____

Write a letter above each number to solve the riddle.

___ ___ ___ ___ ___ ___ ___ ___
 7   6   8   5   4   9   3   12

___ ___ ___!
 5   8   10

Game 4
WHICH IS IT? < > =

Less Than, Greater Than, Equal To

1. Players look at every row on their game boards and determine where they should place *less than*, *greater than*, or *equal to* symbols.
2. The first player draws a symbol card from the bag and places it in one of the boxes to make a true equation (there may be several correct options).
3. If the player cannot place the card to make a true equation, he or she puts the card back into the bag and the next player takes a turn.

Give each player a game board.

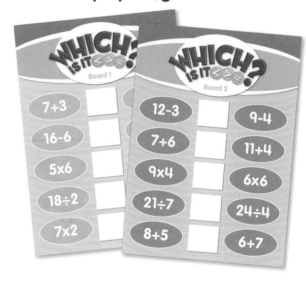

Put the cards into a bag.

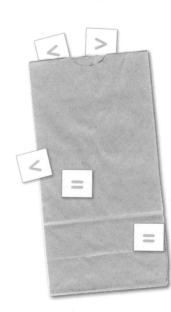

Answer key

Win

1. The first player to make five true equations shouts out, "I win!"
2. Players check the answer key.
3. If the player has correctly placed all of the cards, he or she wins!

Board 1

7+3		4+8
16-6		14-4
5x6		4x9
18÷2		12÷4
7x2		5x3

Math Games, Level D
© Evan-Moor Corp.

WHICH IS IT < > = ?
Board 2

12-3	9-4
7+6	11+4
9×4	6×6
21÷7	24÷4
8+5	6+7

Math Games, Level D
© Evan-Moor Corp.

Board 3

6×3		5×4
14÷2		10÷2
17−6		15−3
17+2		10+9
36÷6		48÷6

Math Games, Level D
© Evan-Moor Corp.

Board 4

42÷6	35÷5
10×3	8×4
12−5	11−6
13+2	9+7
24−8	20−3

Math Games, Level D
© Evan-Moor Corp.

WHICH IS IT? < > =

Board 5

12+2	15+1
6×4	7×3
9−5	11−5
16÷2	24÷3
3×4	2×6

Math Games, Level D
© Evan-Moor Corp.

Board 6

14÷2		20÷4
10+7		8+9
9−3		12−4
7×6		6×6
13+6		12+8

Math Games, Level D
© Evan-Moor Corp.

Math Games • Level D
© Evan-Moor Corp.

 Math Games • Level D
© Evan-Moor Corp.

 Math Games • Level D
© Evan-Moor Corp.

 Math Games • Level D
© Evan-Moor Corp.

 Math Games • Level D
© Evan-Moor Corp.

 Math Games • Level D
© Evan-Moor Corp.

 Math Games • Level D
© Evan-Moor Corp.

 Math Games • Level D
© Evan-Moor Corp.

 Math Games • Level D
© Evan-Moor Corp.

 Math Games • Level D
© Evan-Moor Corp.

 Math Games • Level D
© Evan-Moor Corp.

 Math Games • Level D
© Evan-Moor Corp.

 Math Games • Level D
© Evan-Moor Corp.

 Math Games • Level D
© Evan-Moor Corp.

 Math Games • Level D
© Evan-Moor Corp.

 Math Games • Level D
© Evan-Moor Corp.

 Math Games • Level D
© Evan-Moor Corp.

 Math Games • Level D
© Evan-Moor Corp.

 Math Games • Level D
© Evan-Moor Corp.

 Math Games • Level D
© Evan-Moor Corp.

 Math Games • Level D
© Evan-Moor Corp.

 Math Games • Level D
© Evan-Moor Corp.

 Math Games • Level D
© Evan-Moor Corp.

 Math Games • Level D
© Evan-Moor Corp.

 Math Games • Level D
© Evan-Moor Corp.

 Math Games • Level D
© Evan-Moor Corp.

 Math Games • Level D
© Evan-Moor Corp.

 Math Games • Level D
© Evan-Moor Corp.

 Math Games • Level D
© Evan-Moor Corp.

 Math Games • Level D
© Evan-Moor Corp.

Answer Key

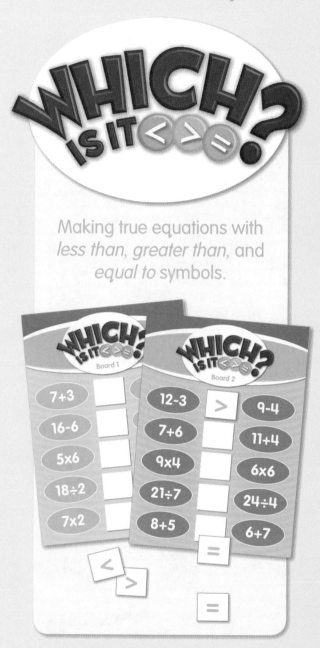

Making true equations with *less than*, *greater than*, and *equal to* symbols.

How to Check:

1. Find the picture that shows your game board.

2. Check to see if the cards between the problems match the cards on your game board.

3. If they do, you win!

Math Games, Level D
EMC 3032 • © Evan-Moor Corp.

Game 4

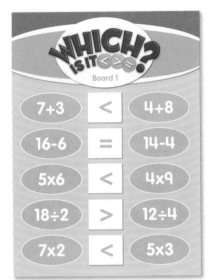
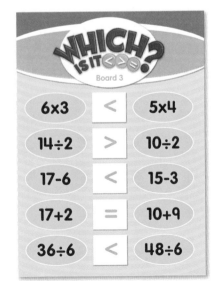
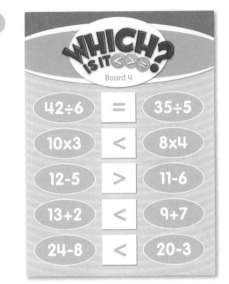
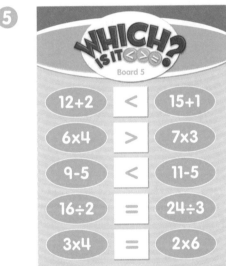

Name _____

Math Games Activity
Less Than, Greater Than, Equal To

Less, Greater, or Equal?

Write <, >, or =.

1. 12−3 ☐ 16−5

2. 3×4 ☐ 2×6

3. 48÷6 ☐ 54÷9

4. 4×7 ☐ 12×5

5. 17+3 ☐ 12+6

6. 20÷4 ☐ 45÷9

7. 3×6 ☐ 2×12

8. 16−11 ☐ 14−7

9. 72÷8 ☐ 49÷7

Name _____

Math Games Activity
Less Than, Greater Than, Equal To

Figure It Out

Complete the equation to make it true.

1 $4 + \square = 5 + 5$

2 $6 + \square > 9 + 2$

3 $6 \times \square > 4 \times 4$

4 $11 \times \square < 8 \times 5$

5 $15 - \square < 18 - 4$

6 $23 - \square = 30 - 10$

7 $27 \div \square = 18 \div 2$

8 $64 \div \square < 81 \div 9$

9 $7 + \square = 5 + 10$

10 $12 + \square = 8 + 8$

Game 5
Ping Pang Pow $
Money: Equal Amounts

Play

1. Assign three players to each set.
2. The first player picks a card from bag 1, adds up the amount shown, and says it out loud. Then the player places it in box 1 on the board and says, "Ping."
3. The next player picks a card from bag 2. If the coins and/or bills add up to the amount in box 1, the player places the card in box 2 and says, "Pang." Then the player goes to step 5.
4. If the coins and/or bills do <u>not</u> add up to the amount in box 1, the player puts the card back into bag 2, and the next player draws a card from bag 2. Play continues until a match is made.
5. The player who makes the match in box 2 draws a card from bag 3. If the coins and/or bills add up to the amount in box 1, the player places the card in box 3 and says, "Pow! I win!"
6. If the coins and/or bills do <u>not</u> add up to the amount in box 1, the player puts the card back into bag 3, and play continues until a player draws the winning card.
7. The winner checks the answer key. If the matches are correct, he or she keeps the cards and a new game begins.

Set A players need:
- Game board
- 21 cards
- 3 brown paper bags

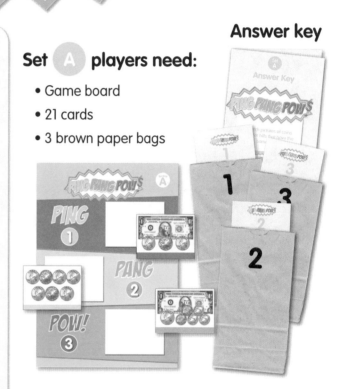

Set B players need:
- Game board
- 21 cards
- 3 brown paper bags

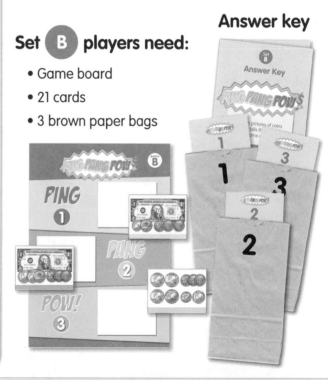

Board A

PING ①

PANG ②

POW! ③

Math Games, Level D
© Evan-Moor Corp.

 Board B

PING 1

PANG 2

POW! 3

Math Games, Level D
© Evan-Moor Corp.

Set 1
Set A
Math Games, Level D
© Evan-Moor Corp.

Set 2
Set A
Math Games, Level D
© Evan-Moor Corp.

Set 3
Set A
Math Games, Level D
© Evan-Moor Corp.

Set 1
Set A
Math Games, Level D
© Evan-Moor Corp.

Set 2
Set A
Math Games, Level D
© Evan-Moor Corp.

Set 3
Set A
Math Games, Level D
© Evan-Moor Corp.

Set 1
Set A
Math Games, Level D
© Evan-Moor Corp.

Set 2
Set A
Math Games, Level D
© Evan-Moor Corp.

Set 3
Set A
Math Games, Level D
© Evan-Moor Corp.

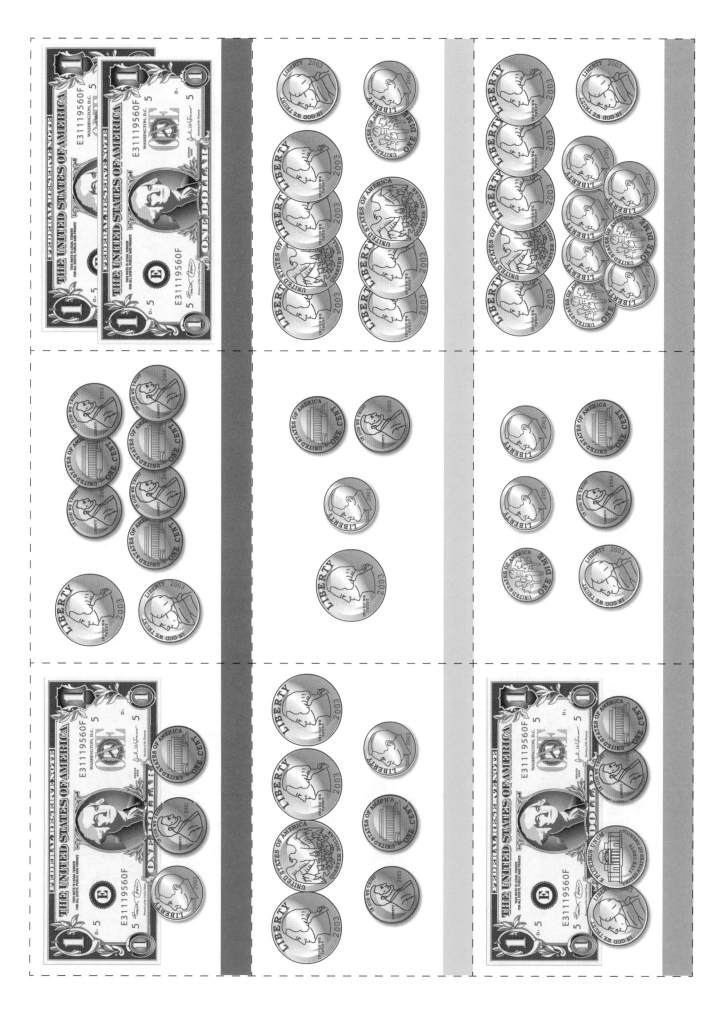

Set A

1

Math Games, Level D
© Evan-Moor Corp.

Set A

2

Math Games, Level D
© Evan-Moor Corp.

Set A

3

Math Games, Level D
© Evan-Moor Corp.

Set A

1

Math Games, Level D
© Evan-Moor Corp.

Set A

2

Math Games, Level D
© Evan-Moor Corp.

Set A

3

Math Games, Level D
© Evan-Moor Corp.

Set A

1

Math Games, Level D
© Evan-Moor Corp.

Set A

2

Math Games, Level D
© Evan-Moor Corp.

Set A

3

Math Games, Level D
© Evan-Moor Corp.

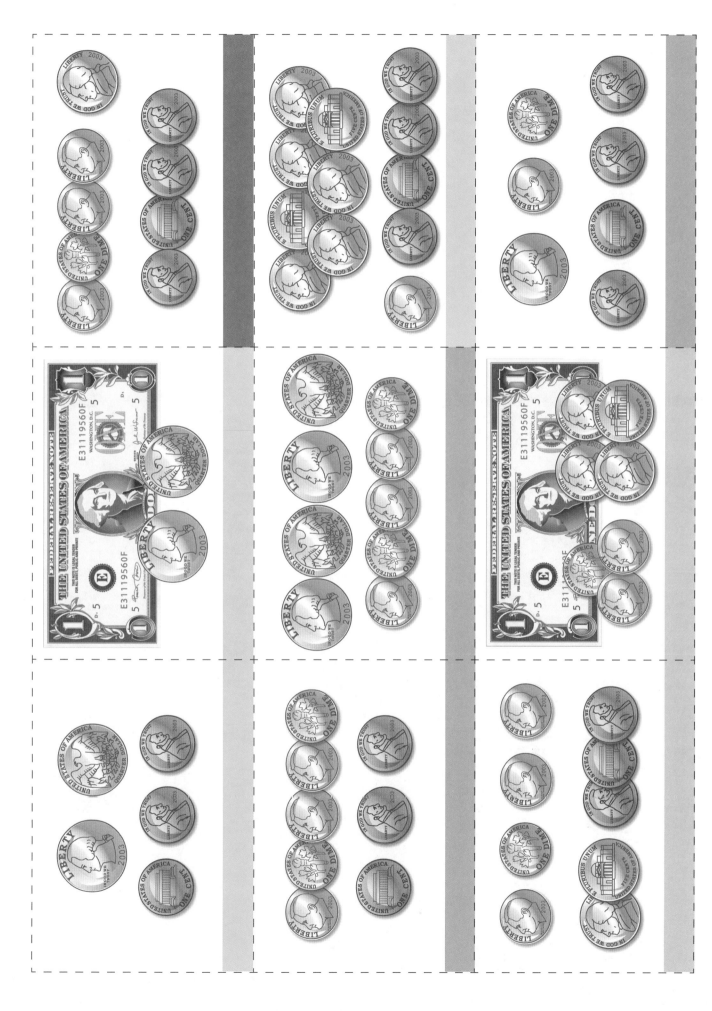

1

Set A

Math Games, Level D
© Evan-Moor Corp.

2

Set A

Math Games, Level D
© Evan-Moor Corp.

3

Set A

Math Games, Level D
© Evan-Moor Corp.

1

Set B

Math Games, Level D
© Evan-Moor Corp.

2

Set B

Math Games, Level D
© Evan-Moor Corp.

3

Set B

Math Games, Level D
© Evan-Moor Corp.

1

Set B

Math Games, Level D
© Evan-Moor Corp.

2

Set B

Math Games, Level D
© Evan-Moor Corp.

3

Set B

Math Games, Level D
© Evan-Moor Corp.

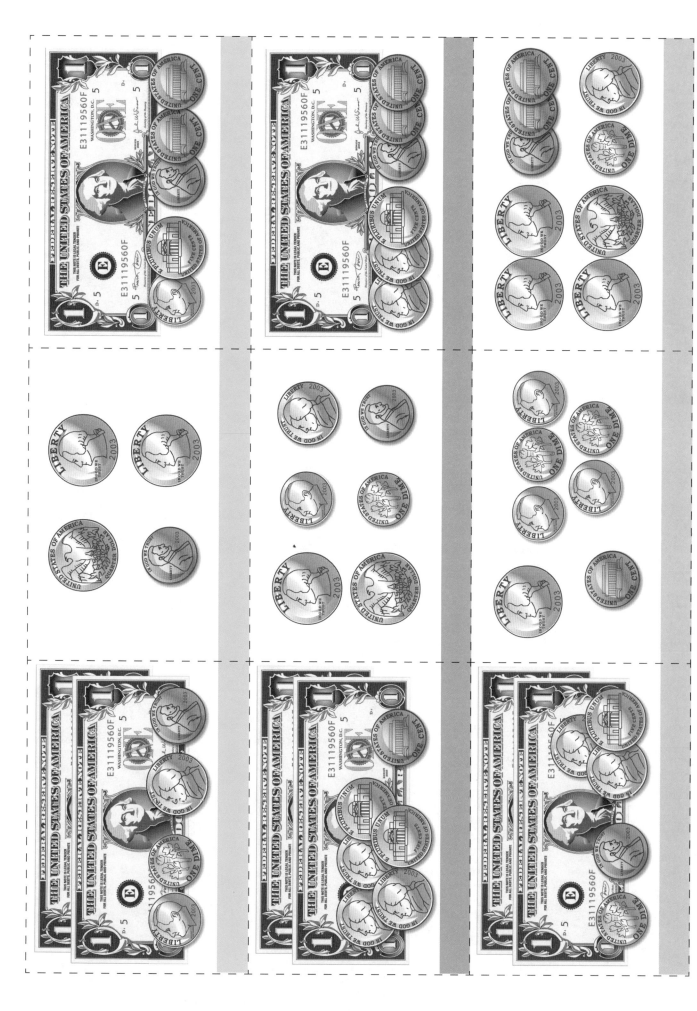

1

Set B

Math Games, Level D
© Evan-Moor Corp.

2

Set B

Math Games, Level D
© Evan-Moor Corp.

3

Set B

Math Games, Level D
© Evan-Moor Corp.

1

Set B

Math Games, Level D
© Evan-Moor Corp.

2

Set B

Math Games, Level D
© Evan-Moor Corp.

3

Set B

Math Games, Level D
© Evan-Moor Corp.

1

Set B

Math Games, Level D
© Evan-Moor Corp.

2

Set B

Math Games, Level D
© Evan-Moor Corp.

3

Set B

Math Games, Level D
© Evan-Moor Corp.

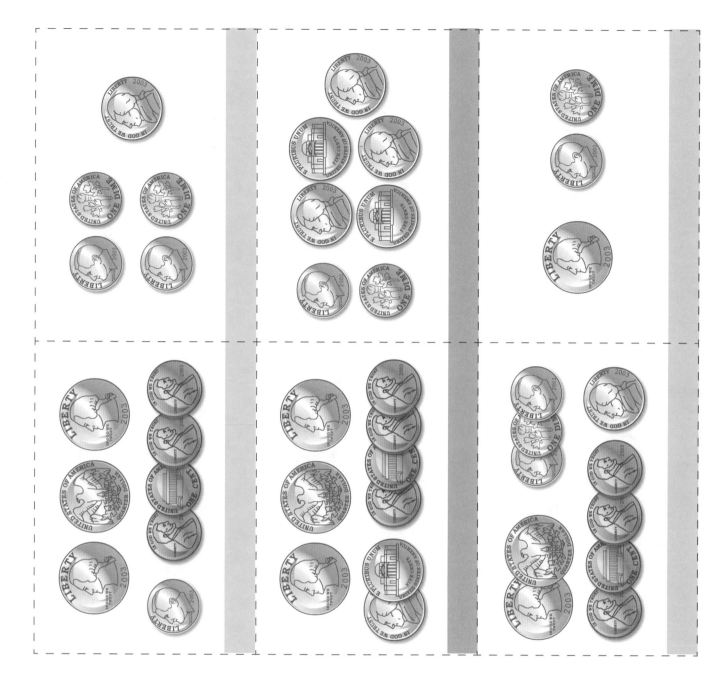

1

Set B

Math Games, Level D
© Evan-Moor Corp.

2

Set B

Math Games, Level D
© Evan-Moor Corp.

3

Set B

Math Games, Level D
© Evan-Moor Corp.

1

Set B

Math Games, Level D
© Evan-Moor Corp.

2

Set B

Math Games, Level D
© Evan-Moor Corp.

3

Set B

Math Games, Level D
© Evan-Moor Corp.

Answer Key

Match pictures of coins and/or bills that have the same value.

How to Check:

1. Look at the first card on your game board. Find it on the answer key.

2. Check to see if the cards next to it match the cards on your game board.

3. If they do, you win!

Answer Key

Match pictures of coins and/or bills that have the same value.

How to Check:

1. Look at the first card on your game board. Find it on the answer key.

2. Check to see if the cards next to it match the cards on your game board.

3. If they do, you win!

Set B

	Ping	Pang	Pow!
45¢			
53¢			
76¢			
89¢			
$1.18			
$1.50			
$2.26			

Set A

	Ping	Pang	Pow!
37¢			
49¢			
64¢			
91¢			
$1.12			
$1.75			
$2.00			

Math Games, Level D
EMC 3032 • © Evan-Moor Corp.

Name _____

Math Games Activity
Money: Equal Amounts

Less Than, Greater Than, Equal to

Use the symbols to compare sums of money.

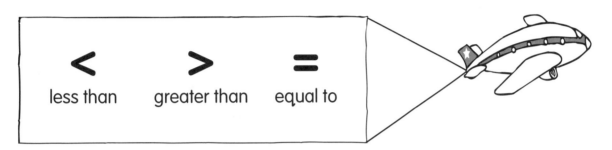

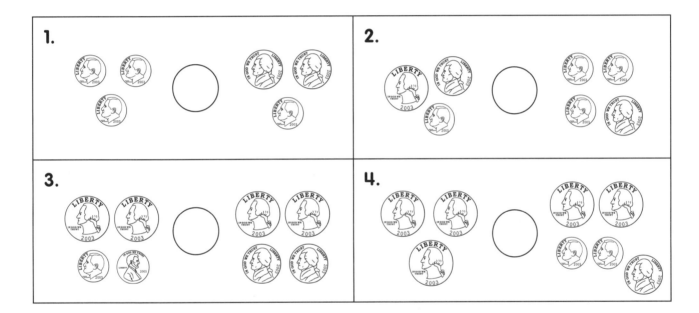

5. 3 nickels ◯ 2 dimes

6. 1 quarter ◯ 5 nickels

7. 15 pennies ◯ 5 nickels

8. 1 dollar ◯ 10 dimes

9. 5 nickels ◯ 3 quarters

10. 3 dimes ◯ 3 quarters

11. 9 dimes ◯ 4 quarters

12. 2 quarters ◯ 6 nickels

How Many Ways Can You Make $2.00?

Use this money table to help you make nine combinations of coins that add up to exactly $2.00.

Penny	Nickel	Dime	Quarter	Half-Dollar	$
200					= $2.00
	5		3	2	= $2.00

Game 6

What Time Is It?

Time to the Minute

Play

1. The first player picks a card from the bag and reads the time out loud.
2. If the clock showing that time is on the player's game board, the player places the card below the clock and draws again.
3. If the clock showing that time already has a card below it, or the clock showing that time is <u>not</u> on the player's game board, the player puts the card back into the bag.
4. The next player takes a turn.

Win

1. The first player to place the correct card below all six clocks calls out, "I win!"
2. Players check the answer key to see if the cards are correctly placed.
3. If the player who called out "I win!" has correctly placed the cards, he or she wins!

Give each player a game board.

Put the cards into a bag.

Answer key

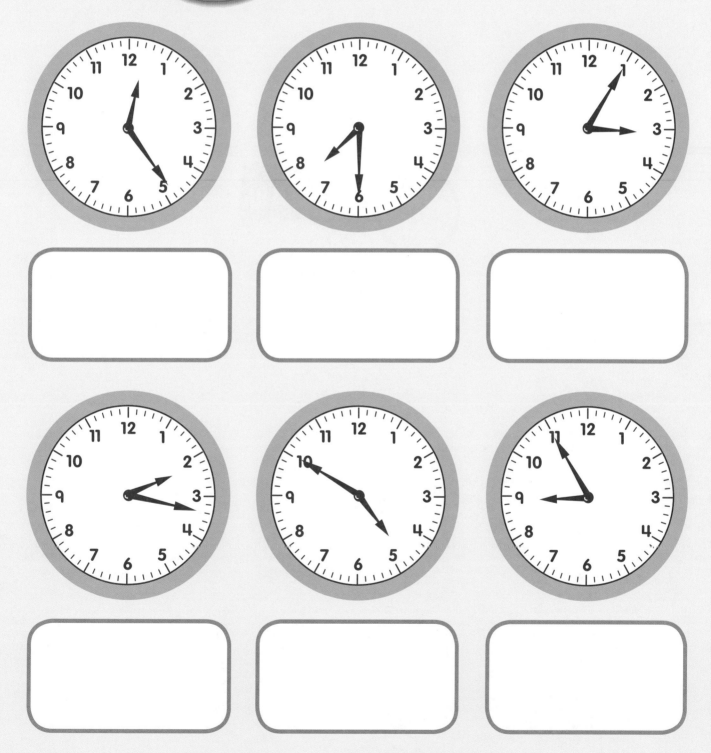

Math Games, Level D
© Evan-Moor Corp.

Math Games, Level D
© Evan-Moor Corp.

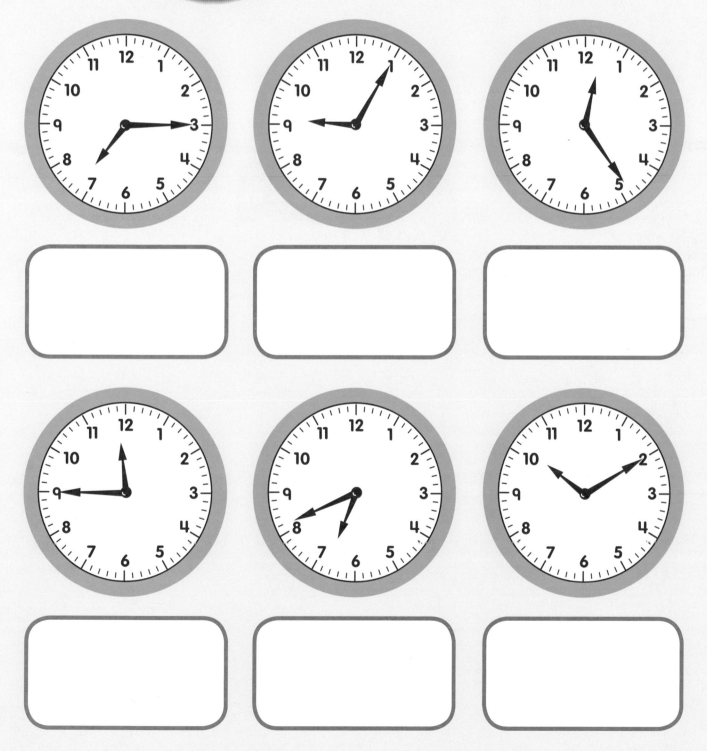

Math Games, Level D
© Evan-Moor Corp.

Math Games, Level D
© Evan-Moor Corp.

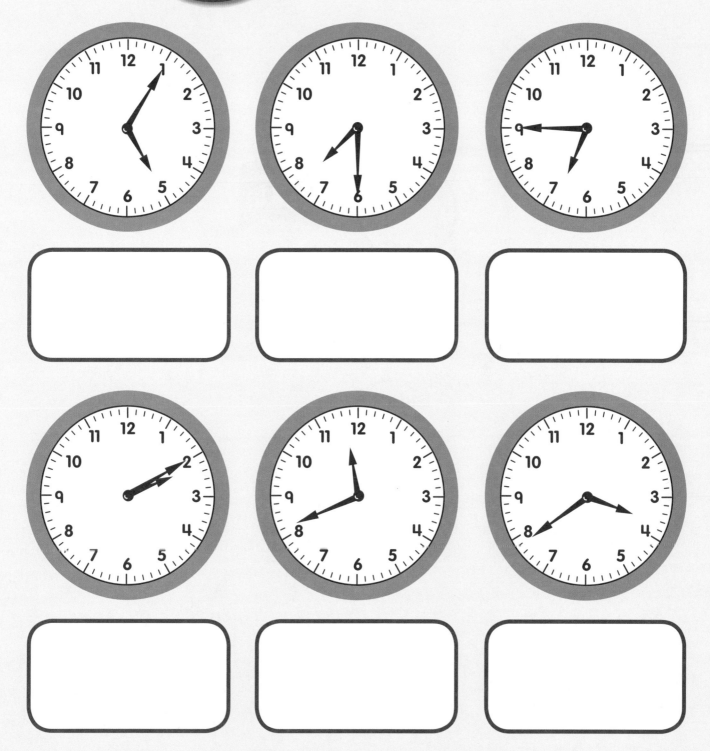

Math Games, Level D
© Evan-Moor Corp.

What Time Is It?

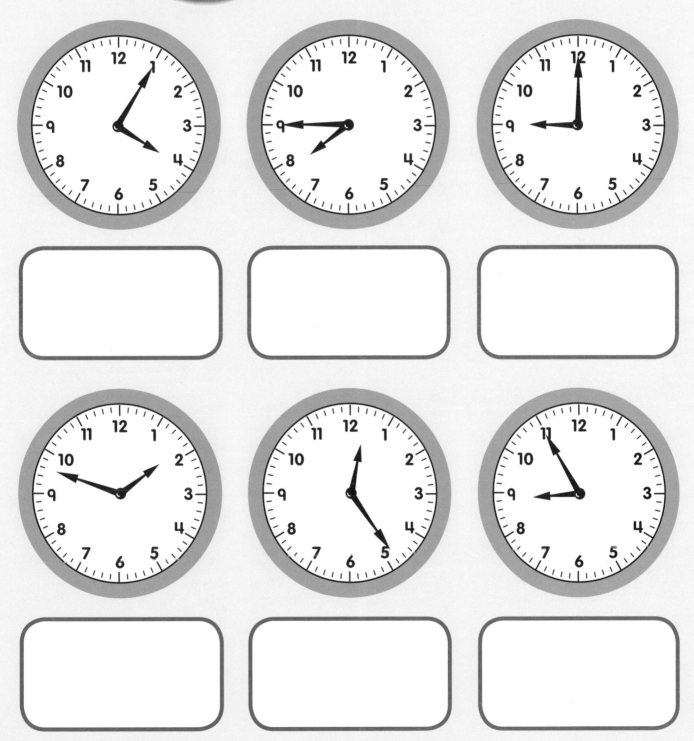

Math Games, Level D
© Evan-Moor Corp.

12:24	7:30	3:05
2:17	4:50	8:55
3:39	6:05	1:48
8:40	5:15	9:30
7:15	9:05	12:24
11:45	6:41	10:10
3:47	8:40	5:15

Math Games, Level D
© Evan-Moor Corp.

Math Games, Level D
© Evan-Moor Corp.

Math Games, Level D
© Evan-Moor Corp.

Math Games, Level D
© Evan-Moor Corp.

Math Games, Level D
© Evan-Moor Corp.

Math Games, Level D
© Evan-Moor Corp.

Math Games, Level D
© Evan-Moor Corp.

Math Games, Level D
© Evan-Moor Corp.

Math Games, Level D
© Evan-Moor Corp.

Math Games, Level D
© Evan-Moor Corp.

Math Games, Level D
© Evan-Moor Corp.

Math Games, Level D
© Evan-Moor Corp.

Math Games, Level D
© Evan-Moor Corp.

Math Games, Level D
© Evan-Moor Corp.

Math Games, Level D
© Evan-Moor Corp.

Math Games, Level D
© Evan-Moor Corp.

Math Games, Level D
© Evan-Moor Corp.

Math Games, Level D
© Evan-Moor Corp.

Math Games, Level D
© Evan-Moor Corp.

Math Games, Level D
© Evan-Moor Corp.

Math Games, Level D
© Evan-Moor Corp.

8:05	4:30	2:45
5:05	7:30	6:45
2:10	11:41	3:39
4:05	7:45	9:00
1:48	12:24	8:55
3:39	8:55	12:24
1:48	4:33	5:05

Math Games, Level D
© Evan-Moor Corp.

Math Games, Level D
© Evan-Moor Corp.

Math Games, Level D
© Evan-Moor Corp.

Math Games, Level D
© Evan-Moor Corp.

Math Games, Level D
© Evan-Moor Corp.

Math Games, Level D
© Evan-Moor Corp.

Math Games, Level D
© Evan-Moor Corp.

Math Games, Level D
© Evan-Moor Corp.

Math Games, Level D
© Evan-Moor Corp.

Math Games, Level D
© Evan-Moor Corp.

Math Games, Level D
© Evan-Moor Corp.

Math Games, Level D
© Evan-Moor Corp.

Math Games, Level D
© Evan-Moor Corp.

Math Games, Level D
© Evan-Moor Corp.

Math Games, Level D
© Evan-Moor Corp.

Math Games, Level D
© Evan-Moor Corp.

Math Games, Level D
© Evan-Moor Corp.

Math Games, Level D
© Evan-Moor Corp.

Math Games, Level D
© Evan-Moor Corp.

Math Games, Level D
© Evan-Moor Corp.

Math Games, Level D
© Evan-Moor Corp.

Answer Key

Match each time card to the correct clock.

How to Check:

1. Find the picture of your game board.

2. Look at the first clock. Check to see if the card below the clock matches the card on your game board. If it does, it's correct!

3. Check all of the cards on your game board.

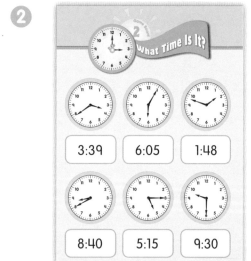

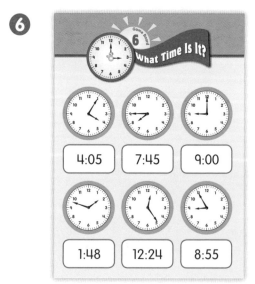

Name _____

Math Games Activity
Time to the Minute

Minute by Minute

Read the clock. Write the time.

1.

  _____ : _____

2.

  _____ : _____

3.

  _____ : _____

4.

  _____ : _____

5.

  _____ : _____

6.

  _____ : _____

7.

  _____ : _____

8.

  _____ : _____

9.

  _____ : _____

Math Games, Level D • EMC 3032 • © Evan-Moor Corp.

What Time Is It? • Game 6

Name _____

Math Games Activity
Time to the Minute

Time Flies

Read the time on the clock. Write the time it was 10 minutes before.
Then write the time it will be 6 minutes later.

10 minutes before		6 minutes after
2:35		2:51
_____ : _____		_____ : _____
_____ : _____		_____ : _____
_____ : _____		_____ : _____
_____ : _____		_____ : _____
_____ : _____		_____ : _____
_____ : _____		_____ : _____

Game 7

Concentration
Equivalent Fractions

Play

1. Place the cards facedown.
2. The first player turns over two cards. If the cards match, the player keeps the pair of cards and plays again.
3. If the cards do <u>not</u> match, the player turns the cards over and the next player takes a turn.

Win

1. Play continues until all of the cards are matched.
2. Then each player looks at the answer key to make sure his or her cards are correctly matched.
3. The player with the most pairs wins!

Players need:

- 26 cards

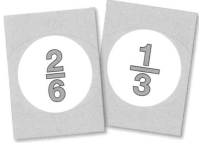

Answer key

$\frac{1}{2}$	$\frac{2}{4}$	$\frac{1}{3}$
$\frac{2}{6}$	$\frac{2}{3}$	$\frac{4}{6}$
$\frac{1}{4}$	$\frac{2}{8}$	$\frac{3}{4}$

Game 7

Math Games, Level D
© Evan-Moor Corp.

Game 7

Math Games, Level D
© Evan-Moor Corp.

Game 7

Math Games, Level D
© Evan-Moor Corp.

Game 7

Math Games, Level D
© Evan-Moor Corp.

Game 7

Math Games, Level D
© Evan-Moor Corp.

Game 7

Math Games, Level D
© Evan-Moor Corp.

Game 7

Math Games, Level D
© Evan-Moor Corp.

Game 7

Math Games, Level D
© Evan-Moor Corp.

Game 7

Math Games, Level D
© Evan-Moor Corp.

$\frac{6}{8}$	$\frac{1}{5}$	$\frac{2}{10}$
$\frac{3}{5}$	$\frac{6}{10}$	$\frac{1}{6}$
$\frac{2}{12}$	$\frac{2}{7}$	$\frac{4}{14}$

Game 7

Math Games, Level D
© Evan-Moor Corp.

Game 7

Math Games, Level D
© Evan-Moor Corp.

Game 7

Math Games, Level D
© Evan-Moor Corp.

Game 7

Math Games, Level D
© Evan-Moor Corp.

Game 7

Math Games, Level D
© Evan-Moor Corp.

Game 7

Math Games, Level D
© Evan-Moor Corp.

Game 7

Math Games, Level D
© Evan-Moor Corp.

Game 7

Math Games, Level D
© Evan-Moor Corp.

Game 7

Math Games, Level D
© Evan-Moor Corp.

3/7	6/14	5/8
10/16	1/8	2/16
1/7	2/14	

Game 7

Math Games, Level D
© Evan-Moor Corp.

Game 7

Math Games, Level D
© Evan-Moor Corp.

Game 7

Math Games, Level D
© Evan-Moor Corp.

Game 7

Math Games, Level D
© Evan-Moor Corp.

Game 7

Math Games, Level D
© Evan-Moor Corp.

Game 7

Math Games, Level D
© Evan-Moor Corp.

Game 7

Math Games, Level D
© Evan-Moor Corp.

Game 7

Math Games, Level D
© Evan-Moor Corp.

Answer Key

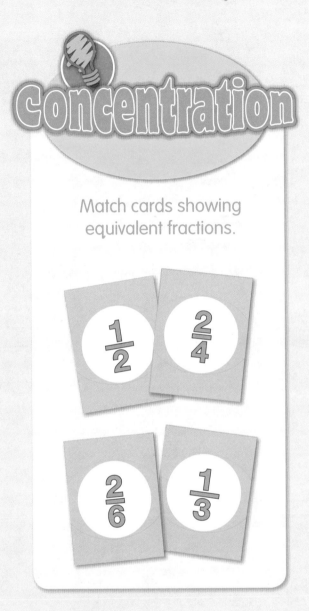

How to Check:

1. Look at one of your pairs. Then find one of the cards in your pair on the answer key.

2. See if the card next to it matches your other card. If it does, you made a pair.

3. Check your other pairs.

$\frac{1}{2} = \frac{2}{4}$	$\frac{1}{6} = \frac{2}{12}$
$\frac{1}{3} = \frac{2}{6}$	$\frac{2}{7} = \frac{4}{14}$
$\frac{2}{3} = \frac{4}{6}$	$\frac{3}{7} = \frac{6}{14}$
$\frac{1}{4} = \frac{2}{8}$	$\frac{5}{8} = \frac{10}{16}$
$\frac{3}{4} = \frac{6}{8}$	$\frac{1}{8} = \frac{2}{16}$
$\frac{1}{5} = \frac{2}{10}$	$\frac{1}{7} = \frac{2}{14}$
$\frac{3}{5} = \frac{6}{10}$	

Name _____

Math Games Activity
Equivalent Fractions

Fraction Buddies

Draw a line to match equivalent fractions.

1. $\frac{2}{7}$ • • $\frac{6}{8}$

2. $\frac{3}{4}$ • • $\frac{4}{14}$

3. $\frac{1}{5}$ • • $\frac{2}{12}$

4. $\frac{1}{2}$ • • $\frac{2}{4}$

5. $\frac{1}{6}$ • • $\frac{4}{6}$

6. $\frac{2}{3}$ • • $\frac{2}{10}$

Name _____

Math Games Activity
Equivalent Fractions

Fraction Action

Read the first fraction. Then circle the fraction below it that is equivalent.

1. $\dfrac{1}{8}$

   $\dfrac{2}{8}$ $\dfrac{2}{16}$ $\dfrac{2}{14}$

2. $\dfrac{3}{4}$

   $\dfrac{6}{8}$ $\dfrac{2}{12}$ $\dfrac{6}{14}$

3. $\dfrac{2}{3}$

   $\dfrac{1}{8}$ $\dfrac{2}{14}$ $\dfrac{4}{6}$

4. $\dfrac{1}{5}$

   $\dfrac{2}{8}$ $\dfrac{6}{10}$ $\dfrac{2}{10}$

5. $\dfrac{5}{8}$

   $\dfrac{10}{16}$ $\dfrac{2}{6}$ $\dfrac{2}{10}$

6. $\dfrac{1}{2}$

   $\dfrac{3}{4}$ $\dfrac{2}{4}$ $\dfrac{1}{6}$

7. $\dfrac{1}{7}$

   $\dfrac{2}{14}$ $\dfrac{2}{16}$ $\dfrac{4}{14}$

8. $\dfrac{1}{3}$

   $\dfrac{2}{4}$ $\dfrac{4}{6}$ $\dfrac{2}{6}$

9. $\dfrac{3}{7}$

   $\dfrac{5}{8}$ $\dfrac{6}{14}$ $\dfrac{1}{7}$

Math Games Answer Key

Page 21
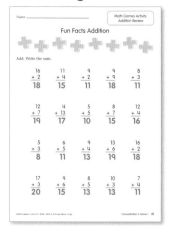

Page 22

Page 43

Page 44
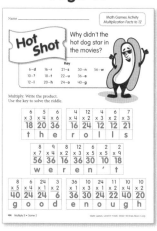

Page 65

Page 66

Page 87
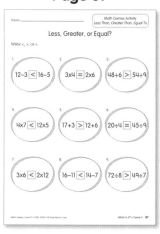

Page 88
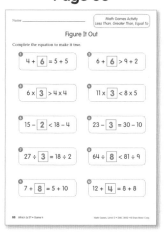

Page 107
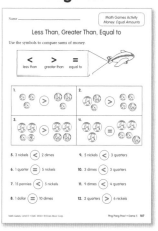

Page 108

Page 129
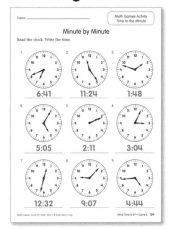

Page 130
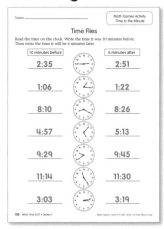

Page 141
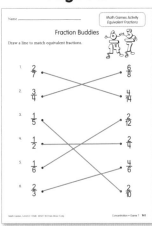

Page 142
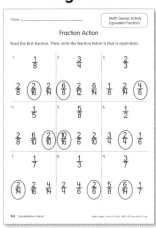